Periodic Table of the Elements

Atomic, Physical, Chemical Properties and Natural Isotopes

Periodic Table

Quick Study ACADEMIC

Legend:

- **Electronegativity:** Pauling scale; measures ability of atom to attract electrons in a chemical bond
- **Atomic Radius:** given in "pm"; 1 pm = 1×10^{-12} m
- **Ionic Radius:** given in "pm"; 1 pm = 1×10^{-10} m
- **Electron Affinity:** energy released in the formation of an anion: given in "eV"
- **1st Ionization Potential:** energy required to remove one electron, forming a cation; given in "eV"
- **Atomic Number:** number of protons
- **Atomic Weight:** weighted average of atomic masses of natural isotopes
  ✧ - mass number of the most stable isotope for each radioactive element

Group numbers: 1, 2, 3, 4, 5, 6, 7, 8, 9, 10, 11, 12, 13, 14, 15, 16, 17, 18

Period 1
- **1 H** 1.008 — Hydrogen — $1s^1$ — Oxidation States: 1 — Electroneg. 2.2 — Atomic Radius 37 — Ionic Radius 2.2 — Electron Affinity 0.75 — 1st Ion. Pot. 13.60
- **2 He** 4.003 — Helium — $1s^2$ — Atomic Radius 2.04 — 1st Ion. Pot. 24.59

Period 2
- **3 Li** 6.94 — Lithium — $1s^2 2s^1$ — Ox. 1 — Electroneg. 0.98 — At. Radius 152 — Ionic Radius (+1)76 — Electron Affinity 0.62 — 1st Ion. Pot. 5.39
- **4 Be** 9.01 — Beryllium — $1s^2 2s^2$ — Ox. 2 — Electroneg. 1.57 — At. Radius 111 — Ionic Radius (+2)45 — 1st Ion. Pot. 9.32
- **5 B** 10.81 — Boron — $1s^2 2s^2 p^1$ — Ox. 3 — Electroneg. 2.04 — At. Radius 80 — Ionic Radius (+3)54 — Electron Affinity 0.28 — 1st Ion. Pot. 8.30
- **6 C** 12.01 — Carbon — $1s^2 2s^2 p^2$ — Ox. ±4,2 — Electroneg. 2.55 — At. Radius 77 — Ionic Radius (+4)16 — Electron Affinity 1.26 — 1st Ion. Pot. 11.26
- **7 N** 14.01 — Nitrogen — $1s^2 2s^2 p^3$ — Ox. ±3,5,4,2 — Electroneg. 3.04 — At. Radius 74 — Ionic Radius (-3)171 — Unstable Anion 14.53
- **8 O** 16.00 — Oxygen — $1s^2 2s^2 p^4$ — Ox. -2 — Electroneg. 3.44 — At. Radius 74 — Ionic Radius (-2)140 — Electron Affinity 1.46 — 1st Ion. Pot. 13.62
- **9 F** 19.00 — Fluorine — $1s^2 2s^2 p^5$ — Ox. -1 — Electroneg. 3.98 — At. Radius 71 — Ionic Radius (-1)133 — Electron Affinity 3.40 — 1st Ion. Pot. 17.42
- **10 Ne** 20.18 — Neon — $1s^2 2s^2 p^6$ — Atomic Radius 21.56 — 1st Ion. Pot. 21.56

Period 3
- **11 Na** 22.99 — Sodium — $[Ne]3s^1$ — Ox. 1 — Electroneg. 0.93 — At. Radius 186 — Ionic Radius (+1)102 — Electron Affinity 0.55 — 1st Ion. Pot. 5.14
- **12 Mg** 24.31 — Magnesium — $[Ne]3s^2$ — Ox. 2 — Electroneg. 1.31 — At. Radius 160 — Ionic Radius (+2)72 — 1st Ion. Pot. 7.65
- **13 Al** 26.98 — Aluminum — $[Ne]3s^2p^1$ — Ox. 3 — Electroneg. 1.61 — At. Radius 143 — Ionic Radius (+3)54 — Electron Affinity 0.44 — 1st Ion. Pot. 5.99
- **14 Si** 28.09 — Silicon — $[Ne]3s^2p^2$ — Ox. 4 — Electroneg. 1.90 — At. Radius 118 — Ionic Radius (+4)40 — Electron Affinity 1.39 — 1st Ion. Pot. 8.15
- **15 P** 30.97 — Phosphorus — $[Ne]3s^2p^3$ — Ox. ±3,5,4 — Electroneg. 2.19 — At. Radius 110 — Ionic Radius (+5)17 — Electron Affinity 0.75 — 1st Ion. Pot. 10.49
- **16 S** 32.07 — Sulfur — $[Ne]3s^2p^4$ — Ox. ±2,4,6 — Electroneg. 2.58 — At. Radius 103 — Ionic Radius (-2)184 — Electron Affinity 2.08 — 1st Ion. Pot. 10.36
- **17 Cl** 35.45 — Chlorine — $[Ne]3s^2p^5$ — Ox. ±1,3,5,7 — Electroneg. 3.16 — At. Radius 99 — Ionic Radius (-1)181 — Electron Affinity 3.61 — 1st Ion. Pot. 12.97
- **18 Ar** 39.95 — Argon — $[Ne]3s^2p^6$ — Atomic Radius — 1st Ion. Pot. 15.76

Period 4
- **19 K** 39.10 — Potassium — $[Ar]4s^1$ — Ox. 1 — Electroneg. 0.82 — At. Radius 227 — Ionic Radius (+1)151 — Electron Affinity 0.50 — 1st Ion. Pot. 4.34
- **20 Ca** 40.08 — Calcium — $[Ar]4s^2$ — Ox. 2 — Electroneg. 1.00 — At. Radius 197 — Ionic Radius (+2)100 — Electron Affinity 0.04 — 1st Ion. Pot. 6.11
- **21 Sc** 44.96 — Scandium — $[Ar]3d^1 4s^2$ — Ox. 3 — Electroneg. 1.36 — At. Radius 161 — Ionic Radius (+3)75 — Electron Affinity 0.19 — 1st Ion. Pot. 6.56
- **22 Ti** 47.87 — Titanium — $[Ar]3d^2 4s^2$ — Ox. 4 — Electroneg. 1.54 — At. Radius 145 — Ionic Radius (+4)61 — Electron Affinity 0.08 — 1st Ion. Pot. 6.83
- **23 V** 50.94 — Vanadium — $[Ar]3d^3 4s^2$ — Ox. 5,3 — Electroneg. 1.63 — At. Radius 131 — Ionic Radius (+5)54 — Electron Affinity 0.53 — 1st Ion. Pot. 6.75
- **24 Cr** 52.00 — Chromium — $[Ar]3d^5 4s^1$ — Ox. 6,3,2 — Electroneg. 1.66 — At. Radius 125 — Ionic Radius (+3)62 — Electron Affinity 0.67 — 1st Ion. Pot. 6.77
- **25 Mn** 54.94 — Manganese — $[Ar]3d^5 4s^2$ — Ox. 7,6,4,2,3 — Electroneg. 1.55 — At. Radius 137 — Ionic Radius (+2)67 — Unstable Anion 7.43
- **26 Fe** 55.85 — Iron — $[Ar]3d^6 4s^2$ — Ox. 2,3 — Electroneg. 1.83 — At. Radius 124 — Ionic Radius (+3)55 — Electron Affinity 0.151 — 1st Ion. Pot. 7.90
- **27 Co** 58.93 — Cobalt — $[Ar]3d^7 4s^2$ — Ox. 2,3 — Electroneg. 1.88 — At. Radius 125 — Ionic Radius (+2)65 — Electron Affinity 0.66 — 1st Ion. Pot. 7.88
- **28 Ni** 58.69 — Nickel — $[Ar]3d^8 4s^2$ — Ox. 2,3 — Electroneg. 1.91 — At. Radius 125 — Ionic Radius (+2)69 — Electron Affinity 1.16 — 1st Ion. Pot. 7.64
- **29 Cu** 63.55 — Copper — $[Ar]3d^{10} 4s^1$ — Ox. 2,1 — Electroneg. 1.90 — At. Radius 128 — Ionic Radius (+2)73 — Electron Affinity 1.24 — 1st Ion. Pot. 7.73
- **30 Zn** 65.39 — Zinc — $[Ar]3d^{10} 4s^2$ — Ox. 2 — Electroneg. 1.65 — At. Radius 133 — Ionic Radius (+2)74 — 1st Ion. Pot. 9.39
- **31 Ga** 69.72 — Gallium — $[Ar]3d^{10} 4s^2 p^1$ — Ox. 3 — Electroneg. 1.81 — At. Radius 122 — Ionic Radius (+3)62 — Electron Affinity 0.3 — 1st Ion. Pot. 5.10
- **32 Ge** 72.61 — Germanium — $[Ar]3d^{10} 4s^2 p^2$ — Ox. 4 — Electroneg. 2.01 — At. Radius 123 — Ionic Radius (+4)53 — Electron Affinity 1.23 — 1st Ion. Pot. 7.90
- **33 As** 74.92 — Arsenic — $[Ar]3d^{10} 4s^2 p^3$ — Ox. ±3,5 — Electroneg. 2.18 — At. Radius 125 — Ionic Radius (+3)58 — Electron Affinity 0.81 — 1st Ion. Pot. 9.81
- **34 Se** 78.96 — Selenium — $[Ar]3d^{10} 4s^2 p^4$ — Ox. -2,4,6 — Electroneg. 2.55 — At. Radius 116 — Ionic Radius (-2)198 — Electron Affinity 2.02 — 1st Ion. Pot. 9.75
- **35 Br** 79.90 — Bromine — $[Ar]3d^{10} 4s^2 p^5$ — Ox. ±1,5 — Electroneg. 2.96 — At. Radius 114 — Ionic Radius (-1)196 — Electron Affinity 3.36 — 1st Ion. Pot. 11.81
- **36 Kr** 83.80 — Krypton — $[Ar]3d^{10} 4s^2 p^6$ — 1st Ion. Pot. 14.00

Period 5
- **37 Rb** 85.47 — Rubidium — $[Kr]5s^1$ — Ox. 1 — Electroneg. 0.82 — At. Radius 248 — Ionic Radius (+1)161 — Electron Affinity 0.49 — 1st Ion. Pot. 4.18
- **38 Sr** 87.62 — Strontium — $[Kr]5s^2$ — Ox. 2 — Electroneg. 0.95 — At. Radius 215 — Ionic Radius (+2)126 — Electron Affinity 0.11 — 1st Ion. Pot. 5.70
- **39 Y** 88.91 — Yttrium — $[Kr]4d^1 5s^2$ — Ox. 3 — Electroneg. 1.22 — At. Radius 178 — Ionic Radius (+3)102 — Electron Affinity 0.31 — 1st Ion. Pot. 6.22
- **40 Zr** 91.22 — Zirconium — $[Kr]4d^2 5s^2$ — Ox. 4 — Electroneg. 1.33 — At. Radius 159 — Ionic Radius (+4)84 — Electron Affinity 0.43 — 1st Ion. Pot. 6.63
- **41 Nb** 92.91 — Niobium — $[Kr]4d^4 5s^1$ — Ox. 5,3 — Electroneg. 1.60 — At. Radius 143 — Ionic Radius (+5)64 — Electron Affinity 0.90 — 1st Ion. Pot. 6.76
- **42 Mo** 95.94 — Molybdenum — $[Kr]4d^5 5s^1$ — Ox. 6,5,4,3,2 — Electroneg. 2.16 — At. Radius 136 — Ionic Radius (+6)59 — Electron Affinity 0.75 — 1st Ion. Pot. 7.09
- **43 Tc** 98 ✧ — Technetium — $[Kr]4d^5 5s^2$ — Ox. 7 — Electroneg. 2.10 — At. Radius 135 — Electron Affinity 0.55 — 1st Ion. Pot. 7.28
- **44 Ru** 101.1 — Ruthenium — $[Kr]4d^7 5s^1$ — Ox. 2,3,4,6,8 — Electroneg. 2.20 — At. Radius 133 — Ionic Radius (+3)68 — Electron Affinity 1.05 — 1st Ion. Pot. 7.36
- **45 Rh** 102.9 — Rhodium — $[Kr]4d^8 5s^1$ — Ox. 2,3,4 — Electroneg. 2.28 — At. Radius 135 — Ionic Radius (+3)67 — Electron Affinity 1.14 — 1st Ion. Pot. 7.46
- **46 Pd** 106.4 — Palladium — $[Kr]4d^{10}$ — Ox. 2,4 — Electroneg. 2.20 — At. Radius 138 — Ionic Radius (+2)64 — Electron Affinity 0.56 — 1st Ion. Pot. 8.34
- **47 Ag** 107.9 — Silver — $[Kr]4d^{10} 5s^1$ — Ox. 1 — Electroneg. 1.93 — At. Radius 145 — Ionic Radius (+1)115 — Electron Affinity 1.30 — 1st Ion. Pot. 7.58
- **48 Cd** 112.4 — Cadmium — $[Kr]4d^{10} 5s^2$ — Ox. 2 — Electroneg. 1.69 — At. Radius 149 — Ionic Radius (+2)95 — 1st Ion. Pot. 8.99
- **49 In** 114.8 — Indium — $[Kr]4d^{10} 5s^2 p^1$ — Ox. 3 — Electroneg. 1.78 — At. Radius 163 — Ionic Radius (+3)80 — Electron Affinity 0.30 — 1st Ion. Pot. 5.79
- **50 Sn** 118.7 — Tin — $[Kr]4d^{10} 5s^2 p^2$ — Ox. 4,2 — Electroneg. 1.96 — At. Radius 141 — Ionic Radius (+4)45 — Electron Affinity 1.11 — 1st Ion. Pot. 7.34
- **51 Sb** 121.8 — Antimony — $[Kr]4d^{10} 5s^2 p^3$ — Ox. ±3,5 — Electroneg. 2.05 — At. Radius 145 — Ionic Radius (+3)76 — Electron Affinity 1.07 — 1st Ion. Pot. 8.64
- **52 Te** 127.6 — Tellurium — $[Kr]4d^{10} 5s^2 p^4$ — Ox. ±2,4,6 — Electroneg. 2.10 — At. Radius 143 — Ionic Radius (-2)221 — Electron Affinity 1.97 — 1st Ion. Pot. 9.01
- **53 I** 126.9 — Iodine — $[Kr]4d^{10} 5s^2 p^5$ — Ox. ±1,5,7 — Electroneg. 2.66 — At. Radius 133 — Ionic Radius (-1)220 — Electron Affinity 3.06 — 1st Ion. Pot. 10.45
- **54 Xe** 131.3 — Xenon — $[Kr]4d^{10} 5s^2 p^6$ — 1st Ion. Pot. 12.13

Period 6
- **55 Cs** 132.9 — Cesium — $[Xe]6s^1$ — Ox. 1 — Electroneg. 0.79 — At. Radius 265 — Ionic Radius (+1)174 — Electron Affinity 0.47 — 1st Ion. Pot. 3.89
- **56 Ba** 137.3 — Barium — $[Xe]6s^2$ — Ox. 2 — Electroneg. 0.89 — At. Radius 217 — Ionic Radius (+2)142 — Electron Affinity 0.15 — 1st Ion. Pot. 5.21
- **57 La** 138.9 — Lanthanum — $[Xe]5d^1 6s^2$ — Ox. 3 — Electroneg. 1.10 — At. Radius 187 — Ionic Radius (+3)116 — Electron Affinity 0.5 — 1st Ion. Pot. 5.77
- **72 Hf** 178.5 — Hafnium — $[Xe]4f^{14} 5d^2 6s^2$ — Ox. 4 — Electroneg. 1.3 — At. Radius 156 — Ionic Radius (+4)83 — Electron Affinity ~0 — 1st Ion. Pot. 6.83
- **73 Ta** 180.9 — Tantalum — $[Xe]4f^{14} 5d^3 6s^2$ — Ox. 5 — Electroneg. 1.5 — At. Radius 143 — Ionic Radius (+5)64 — Electron Affinity 0.32 — 1st Ion. Pot. 7.89
- **74 W** 183.8 — Tungsten — $[Xe]4f^{14} 5d^4 6s^2$ — Ox. 6,5,4,3,2 — Electroneg. 1.7 — At. Radius 137 — Ionic Radius (+6)60 — Electron Affinity 0.86 — 1st Ion. Pot. 7.98
- **75 Re** 186.2 — Rhenium — $[Xe]4f^{14} 5d^5 6s^2$ — Ox. 7,6,4,2,-1 — Electroneg. 1.9 — At. Radius 137 — Ionic Radius (+7)53 — Electron Affinity 0.15 — 1st Ion. Pot. 7.88
- **76 Os** 190.2 — Osmium — $[Xe]4f^{14} 5d^6 6s^2$ — Ox. 2,3,4,6,8 — Electroneg. 2.2 — At. Radius 134 — Ionic Radius (+4)63 — Electron Affinity 1.10 — 1st Ion. Pot. 8.7
- **77 Ir** 192.2 — Iridium — $[Xe]4f^{14} 5d^7 6s^2$ — Ox. 2,3,4,6 — Electroneg. 2.2 — At. Radius 136 — Ionic Radius (+4)63 — Electron Affinity 1.57 — 1st Ion. Pot. 9.1
- **78 Pt** 195.1 — Platinum — $[Xe]4f^{14} 5d^9 6s^1$ — Ox. 2,4 — Electroneg. 2.2 — At. Radius 139 — Ionic Radius (+4)63 — Electron Affinity 2.13 — 1st Ion. Pot. 9.0
- **79 Au** 197.0 — Gold — $[Xe]4f^{14} 5d^{10} 6s^1$ — Ox. 3,1 — Electroneg. 2.4 — At. Radius 144 — Ionic Radius (+3)85 — Electron Affinity 2.31 — 1st Ion. Pot. 9.23
- **80 Hg** 200.6 — Mercury — $[Xe]4f^{14} 5d^{10} 6s^2$ — Ox. 2,1 — Electroneg. 1.9 — At. Radius 150 — Ionic Radius (+2)102 — 1st Ion. Pot. 10.44
- **81 Tl** 204.4 — Thallium — $[Xe]4f^{14} 5d^{10} 6s^2 p^1$ — Ox. 3,1 — Electroneg. 1.8 — At. Radius 170 — Ionic Radius (+1)159 — Electron Affinity 0.2 — 1st Ion. Pot. 6.11
- **82 Pb** 207.2 — Lead — $[Xe]4f^{14} 5d^{10} 6s^2 p^2$ — Ox. 4,2 — Electroneg. 1.8 — At. Radius 175 — Ionic Radius (+2)119 — Electron Affinity 0.36 — 1st Ion. Pot. 7.42
- **83 Bi** 209.0 — Bismuth — $[Xe]4f^{14} 5d^{10} 6s^2 p^3$ — Ox. 3,5 — Electroneg. 1.9 — At. Radius 155 — Ionic Radius (+3)103 — Electron Affinity 0.95 — 1st Ion. Pot. 7.29
- **84 Po** 209 ✧ — Polonium — $[Xe]4f^{14} 5d^{10} 6s^2 p^4$ — Ox. 2,4 — Electroneg. 2.0 — At. Radius 167 — Ionic Radius (+4)— — Electron Affinity 1.9 — 1st Ion. Pot. 8.42
- **85 At** 210 ✧ — Astatine — $[Xe]4f^{14} 5d^{10} 6s^2 p^5$ — Ox. ±1,3,5,7 — Electroneg. 2.2 — At. Radius — Electron Affinity 2.8
- **86 Rn** 222 ✧ — Radon — $[Xe]4f^{14} 5d^{10} 6s^2 p^6$ — 1st Ion. Pot. 10.75

Period 7
- **87 Fr** 223 ✧ — Francium — $[Rn]7s^1$ — Ox. 1 — Electroneg. 0.7 — At. Radius — Ionic Radius (+1)— — Electron Affinity 0.46
- **88 Ra** 226 ✧ — Radium — $[Rn]7s^2$ — Ox. 2 — Electroneg. 0.9 — At. Radius — Ionic Radius (+2)162 — 1st Ion. Pot. 5.28
- **89 Ac** 227 ✧ — Actinium — $[Rn]6d^1 7s^2$ — Ox. 3 — Electroneg. 1.1 — At. Radius 188 — Ionic Radius (+3)— — 1st Ion. Pot. 5.17
- **104 Rf** 261 ✧ — Rutherfordium — $[Rn]5f^{14} 6d^2 7s^2$ — disc. 1964
- **105 Db** 262 ✧ — Dubnium — $[Rn]5f^{14} 6d^3 7s^2$ — disc. 1967
- **106 Sg** 266 ✧ — Seaborgium — $[Rn]5f^{14} 6d^4 7s^2$ — disc. 1974
- **107 Bh** 264 ✧ — Bohrium — $[Rn]5f^{14} 6d^5 7s^2$ — disc. 1981
- **108 Hs** 269 ✧ — Hassium — $[Rn]5f^{14} 6d^6 7s^2$ — disc. 1984
- **109 Mt** 268 ✧ — Meitnerium — $[Rn]5f^{14} 6d^7 7s^2$ — disc. 1982
- **110 Uun** 271 ✧ — Ununnilium — disc. 1994 — Darmstadt, Germany
- **111 Uuu** 272 ✧ — Unununium — disc. 1994 — Darmstadt, Germany
- **112 Uub** 277 ✧ — Ununbium — disc. 1996 — Darmstadt, Germany
- **113 ?** — Not discovered
- **114 Uuq** 289 ✧ — Ununquadium — disc. 1999 — Dubna, Russia
- **115 ?** — Not discovered
- **116 Uuh** 289 ✧ — Ununhexium — disc. — Berkeley, Californ. — RETRACTED
- **117 ?** — Not discovered
- **118 Uuo** 293 ✧ — Ununoctium — disc. — Berkeley, Californ. — RETRACTED

Labels: ALKALI METALS; ALKALINE EARTH METALS; HALOGENS; NOBLE GASES

Lanthanide Series
- **58 Ce** 140.1 — Cerium — $[Xe]4f^1 5d^1 6s^2$ — Ox. 3,4 — Electroneg. 1.12 — At. Radius 183 — Ionic Radius (+3)114 — 1st Ion. Pot. 5.54
- **59 Pr** 140.9 — Praseodymium — $[Xe]4f^3 6s^2$ — Ox. 3,4 — Electroneg. 1.13 — At. Radius 182 — Ionic Radius (+3)113 — 1st Ion. Pot. 5.46
- **60 Nd** 144.2 — Neodymium — $[Xe]4f^4 6s^2$ — Ox. 3 — Electroneg. 1.14 — At. Radius 181 — Ionic Radius (+3)114 — 1st Ion. Pot. 5.53
- **61 Pm** 145 ✧ — Promethium — $[Xe]4f^5 6s^2$ — Ox. 3 — At. Radius 181 — Ionic Radius (+3)109 — 1st Ion. Pot. 5.55
- **62 Sm** 150.4 — Samarium — $[Xe]4f^6 6s^2$ — Ox. 3,2 — Electroneg. 1.17 — At. Radius 180 — Ionic Radius (+3)108 — 1st Ion. Pot. 5.64
- **63 Eu** 152.0 — Europium — $[Xe]4f^7 6s^2$ — Ox. 3,2 — At. Radius 199 — Ionic Radius (+3)107 — 1st Ion. Pot. 5.67
- **64 Gd** 157.3 — Gadolinium — $[Xe]4f^7 5d^1 6s^2$ — Ox. 3 — Electroneg. 1.20 — At. Radius 179 — Ionic Radius (+3)105 — 1st Ion. Pot. 6.15
- **65 Tb** 158.9 — Terbium — $[Xe]4f^9 6s^2$ — Ox. 3,4 — At. Radius 176 — Ionic Radius (+3)118 — 1st Ion. Pot. 5.86
- **66 Dy** 162.5 — Dysprosium — $[Xe]4f^{10} 6s^2$ — Ox. 3 — Electroneg. 1.22 — At. Radius 175 — Ionic Radius (+3)103 — 1st Ion. Pot. 5.94
- **67 Ho** 164.9 — Holmium — $[Xe]4f^{11} 6s^2$ — Ox. 3 — Electroneg. 1.23 — At. Radius 174 — Ionic Radius (+3)100 — 1st Ion. Pot. 6.02
- **68 Er** 167.3 — Erbium — $[Xe]4f^{12} 6s^2$ — Ox. 3 — Electroneg. 1.24 — At. Radius 173 — Ionic Radius (+3)100 — 1st Ion. Pot. 6.11
- **69 Tm** 168.9 — Thulium — $[Xe]4f^{13} 6s^2$ — Ox. 3,2 — Electroneg. 1.25 — At. Radius 173 — Ionic Radius (+3)99 — 1st Ion. Pot. 6.18
- **70 Yb** 173.0 — Ytterbium — $[Xe]4f^{14} 6s^2$ — Ox. 3,2 — At. Radius 194 — Ionic Radius (+3)99 — 1st Ion. Pot. 6.25
- **71 Lu** 175.0 — Lutetium — $[Xe]4f^{14} 5d^1 6s^2$ — Ox. 3 — Electroneg. — At. Radius 172 — Ionic Radius (+3)98 — 1st Ion. Pot. 5.43

Actinide Series
- **90 Th** 232.0 — Thorium — $[Rn]6d^2 7s^2$ — Ox. 4 — Electroneg. 1.3 — At. Radius 180 — Ionic Radius (+4)94 — 1st Ion. Pot. 6.08
- **91 Pa** 231.0 — Protactinium — $[Rn]5f^2 6d^1 7s^2$ — Ox. 5,4 — Electroneg. 1.5 — At. Radius 161 — Ionic Radius (+5)78 — 1st Ion. Pot. 5.89
- **92 U** 238.0 — Uranium — $[Rn]5f^3 6d^1 7s^2$ — Ox. 6,5,4,3 — Electroneg. 1.7 — At. Radius 139 — Ionic Radius (+6)73 — 1st Ion. Pot. 6.19
- **93 Np** 237 ✧ — Neptunium — $[Rn]5f^4 6d^1 7s^2$ — Ox. 6,5,4,3 — Electroneg. 1.3 — At. Radius 131 — Ionic Radius (+4)86 — 1st Ion. Pot. 6.27
- **94 Pu** 244 ✧ — Plutonium — $[Rn]5f^6 7s^2$ — Ox. 6,5,4,3 — Electroneg. 1.3 — At. Radius 151 — Ionic Radius (+4)86 — 1st Ion. Pot. 6.06
- **95 Am** 243 ✧ — Americium — $[Rn]5f^7 7s^2$ — Ox. 6,5,4,3 — At. Radius 131 — Ionic Radius (+3)98 — 1st Ion. Pot. 5.99
- **96 Cm** 247 ✧ — Curium — $[Rn]5f^7 6d^1 7s^2$ — Ox. 3 — Ionic Radius (+3)97 — 1st Ion. Pot. 6.02
- **97 Bk** 247 ✧ — Berkelium — $[Rn]5f^9 7s^2$ — Ox. 4,3 — Ionic Radius (+3)96 — 1st Ion. Pot. 6.23
- **98 Cf** 251 ✧ — Californium — $[Rn]5f^{10} 7s^2$ — Ox. 3 — Ionic Radius (+3)95 — 1st Ion. Pot. 6.30
- **99 Es** 252 ✧ — Einsteinium — $[Rn]5f^{11} 7s^2$ — Ox. 3 — 1st Ion. Pot. 6.42
- **100 Fm** 257 ✧ — Fermium — $[Rn]5f^{12} 7s^2$ — Ox. 3 — 1st Ion. Pot. 6.50
- **101 Md** 258 ✧ — Mendelevium — $[Rn]5f^{13} 7s^2$ — Ox. 3,2 — 1st Ion. Pot. 6.58
- **102 No** 259 ✧ — Nobelium — $[Rn]5f^{14} 7s^2$ — Ox. 3,2 — 1st Ion. Pot. 6.65
- **103 Lr** 262 ✧ — Lawrencium — $[Rn]5f^{14} 6d^1 7s^2$ — Ox. 3

REFERENCE: CRC Handbook of Chemistry and Physics - 81st edition,
ADVISOR: Mark Jackson, Chemistry Professor - Florida State University
LAYOUT: John Ford, CEO - BarCharts, Inc. ® 2001 BarCharts, Inc.
study.com
$4.95 U.S.
$7.50 CAN
October 2001

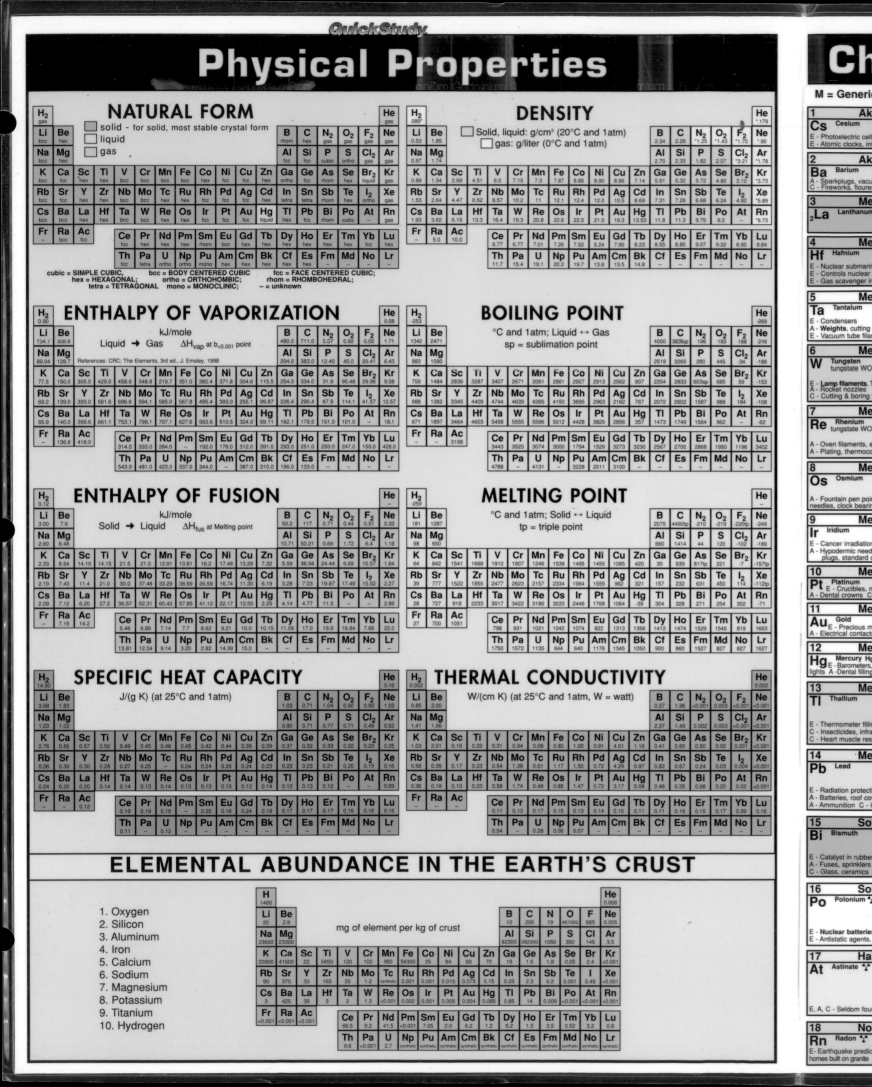

Chemical Properties & Common Uses

M = Generic symbol for element in compound; E = Element raw material form; A = Alloy, blend or mixture; C = Compound; **Bold** = Most Important Use

1 — Akali metals; compounds with M(1+) valences

Cs Cesium
E - Photoelectric cells
E - Atomic clocks, infrared lamps

Rb Rubidium
E - Photoelectric cells
E - Vacuum tubes, heart research

K Potassium
C - Fertilizer, glass, lenses
C - Matches, gunpowder, salt substitute

Na Sodium
A - Street lights A - Nuclear reactor coolant, batteries C - Salt, soda, glass

Li Lithium
E - Pacemaker Batteries A - Lubricant additive, alloys used in space C - Glass & pharmaceuticals

H Hydrogen — $M(1+)$ compounds, acids: metal hydrides $H(1-)$
E - Rocket fuel, hydrogenation of fats
E - Petroleum desulfurization, H_2O, ammonia

2 — Alkaline earth metals; compounds with M(2+) valences

Ba Barium
E - Sparkplugs, vacuum tubes
C - Fireworks, flourescent lamps

Sr Strontium
C - Nuclear batteries in bouys
V - Fireworks, phosphorescent paint

Ca Calcium
C - Metallurgy A-Cable insulation, batteries
E - Fertilizer, concrete, plaster of Paris C - Fireplace bricks, pigments, fillers

Mg Magnesium
E - Airplanes, racing bikes
C - Watch springs, sparkfree tools

Be Beryllium
E - X-ray tube windows
C - Watch springs, sparkfree tools

3 — Metal; compounds and ligand complexes M(3+)

La Lanthanum
C - Color TV screens, radar, lasers
C - Camera lenses, fireproof bricks

Y Yttrium
C - Color TV screens, radar, lasers
E - Seed germinating agents

Sc Scandium
E - Leak detectors, A - Space industry materials C - Seed germinating agents

4 — Metal; compounds: M(2+...4+); ligand complexes: M(1-,0,1+)

Hf Hafnium
E - Nuclear submarines
E - Controls nuclear reactors
E - Gas scavenger in vacuum tubes

Zr Zirconium
E - Nuclear fuel rods, catalytic converters
A - Percussion caps
C - Furnace bricks

Ti Titanium — dioxide TiO_2; metatitanate $TiO_3(2-)$
E - Heat exchanger A - Airplane motors
A - Bone pins C - Pigments for paint/paper

5 — Metal; compounds: M(1+...5+); ligand complexes: M(1-,0,1+)

Ta Tantalum
E - Condensers
A - Weights, cutting tools
E - Vacuum tube filaments

Nb Niobium
A - Cutting tools, pipelines
A - Super magnets, welding rods

V Vanadium — vanadate $VO_4(3-)$
E - Construction materials, tools
E - Springs, jet engines

6 — Metal; compounds: M(1+...6+); ligand complexes: M(2-...1+)

W Tungsten — tungstate $WO_4(2-)$
E - Lamp filaments, TV, welding electrodes
E - Rocket nozzles
C - Cutting & boring tools

Mo Molybdenum — molybdate MoO_5, $MoO_4(2-)$, dimolybdate $Mo_2O_7(2-)$, Mo-cyanate Mo-$CN_6(4-)$
E - Filament in electric heaters
A - Rocket motors C - Lubricant

Cr Chromium — chromate $CrO_4(2-)$, dichromate $Cr_2O_7(2-)$
E - **Plating for car parts**
A - Tools, knives C - Camouflage paint
C - Stereo, video tape, lasers

7 — Metal; compounds: M(1+...7+); ligand complexes: M(3-...1+)

Re Rhenium — rhenate $WO_4(2-)$
A - Oven filaments, electrodes, jewelry
E - Plating, thermocouples

Tc Technetium — synthetic
E - Radiation source for medical research

Mn Manganese — permanganate $MnO_4(1-)$
A - Tools, axles, steel for rail switches
A - Safes, plows, batteries

8 — Metal; compounds: M(1+...8+); ligand complexes: M(2-...1+)

Os Osmium
A - Fountain pen points, compass needles, clock bearings, decorations

Ru Ruthenium
E - Wear treatment, thickness meters for egg shells
A - Fountain pen point, electrical contacts

Fe Iron — Fe-cyanate $Fe(CN)_6(3$-and$4-)$, ferrate $FeO_4(2-)$
E - Bikes, cars, bridges, magnets, machines C - Nails, tools, tin cans

9 — Metal; compounds: M(2+...6+); ligand complexes: M(2-...1+)

Ir Iridium
E - Cancer irradiation
A - Hypodermic needles, helicopter spark plugs, standard one-meter bar

Rh Rhodium — rhenate $ReO_4(1-)$
E - **Headlight reflectors**, telephone
E - Relays, fountain pen points
A - Airplane spark plugs

Co Cobalt — cobaltate $CoO_4(2+)$, Co-cyanate $Co(CN)_6(3$- and $4-)$
E - Gamma radiation A - Permanent magnet C - **Razor blades** C - Catalytic converter

10 — Metal; compounds: M(1+...6+); ligand complexes: M(1-,0,1+)

Pt Platinum
E - Crucibles, nitric acid production
A - Dental crowns C - Anti-tumor agent

Pd Palladium
E - Catalytic converters, hydrogen seperation A - Dental crowns, tel. relays

Ni Nickel
E - Coins A-Knives, forks, spoons
C - White gold C-Rechargeable batteries

11 — Metal; compounds: M(1+...3+); ligand complexes: M(1-,0,1+)

Au Gold
E - Precious metal A - Jewelry
A - Electrical contacts, dental crowns

Ag Silver
E - Mirror, batteries; A - Silverware
C - Photograph film and paper

Cu Copper
E - Cable, wire A - Pennies, bells
A - Bronze sculpture, Statue of Liberty

12 — Metal; compounds and ligand complexes M(2+)

Hg Mercury — $Hg_2(2+)$
E - Barometers, thermometers, street lights A - Dental fillings C - Seed protection

Cd Cadmium
E - Rechargeable batteries A-Nuclear reactor regulator C - Red/yellow pigments

Zn Zinc
E - Anti-corrosion coating, batteries A - Watergas valves C - White rubber pigment

13 — Metal; compounds and ligand complexes: M(1+...3+)

Tl Thallium
E - Thermometer filling
C - Insecticides, infrared detectors
C - Heart muscle research

In Indium
E - Solar cells, mirrors
A - Regulator in nuclear power
C - Photo cells, transistors
C - Blood and lung research

Ga Gallium
E - Quartz thermometers
C - Computer memory, transistors laser
C - Laser diodes, used to locate tumors

Al Aluminum
E - Window frames, doorknobs, cable
E - Cable, foil, fireworks, flashbulbs
A - Cars, rockets, planes

B Boron — metalloid; $B(2+,3,3-)$; compounds: borates (BO_4, BO_3), tetraborates $B_4O_7(2-)$, boranes, borides
C - Tennis rackets, regulator in nuclear plants, heat resistant glass, eye disinfectant

14 — Metal; compounds and ligand complexes: M(2+,4+)

Pb Lead — metal
E - Radiation protection
E - Batteries, roof coverings, solders
A - Ammunition C - Gas additives

Sn Tin — metal; hydroxy stannate $Sn(OH)_6(-2)$
E - Cups, plates, coins, organ pipes
C - Opalescent glass, enamel

Ge Germanium — metalloid; germanate $GeO_2(2-)$, $GeO_3(2-)$
E - Infrared prisms, reflector in projectors
E - Wide angle lenses A - Dentistry

Si Silicon — metalloid; silicates $SiO_4(4-)$ silanes (Si-hydrides) $Si(4-)$ silicides silicic acid H_4SiO_4
E - Microchips, solar cells C - Tools
E - Quartz, cement, glass, grease, oils
C - Pencils, diamonds, steel, controls nuclear reaction
E - Tire coolant C - Plastics, life

C Carbon — $C(2+,4+,4-)$ carbonate $CO_3(2-)$, bicarbonate $HCO_3(1-)$, acetate $C_2H_3O_2(1-)$, oxalate $C_2O_4(2-)$, carbides $(4-)$

15 — Solid; compounds: M(3+,5+,3-)

Bi Bismuth
E - Catalyst in rubber production
A - Fuses, sprinklers
C - Glass, ceramics

Sb Antimony
A - Solder, bearings, lead batteries
A - Mascara, infrared detectors

As Arsenic — arsenide $(3-)$
E - Shotgun pellets
A - Metal for mirrors
C - Glass, lasers
C - Light-emitting diodes=LED

P Phosphorus — nitride $N(3-)$, phosphide $(3-)$, phosphate $(PO_4,-3)$, phosphoric acid H_3PO_4
E - Fireworks, matches C - Pesticides
C - Fertilizer, detergents, toothpaste

N Nitrogen — N_2 gas; $N(3-...5+)$; ammonium $NH_4(1+)$; nitrite $NO_2(1-)$ nitrate $NO_3(1-)$ oxides - NOx; azide $N_3(1-)$ diimide $N_2(2-)$ amide $NH_2(1-)$; acid: nitric HNO_3, nitrous HNO_2
E - Cryogenic surgery, liquid coolant, ammonia production

16 — Solid; compounds: M(2+,4+,6+,2-)

Po Polonium — metal
E - **Nuclear batteries**, neutron source
E - Antistatic agents, film cleaner

Te Tellurium — metalloid; tellurate $TeO_4(2-)$
E - **Vulcanization of rubber**, percussion caps, battery plate protector
A - Electric resistors

Se Selenium — selenides $Se(2-)$, selenate $SeO_4(2-)$; acids of SeO_4
E - **Light meters**, copy machines, solar cells
C - Anti-dandruff shampoo

S Sulfur — solid; sulfide $S_2(1-)$, sulfate $SO_4(2-)$, sulfite $SO_3(2-)$; thiosulfate $S_2O_3(2-)$ thio...(S added or subst. from above); acids: sulfuric H_2SO_4, sulfurous H_2SO_3
E - Combustion, steel production
E - Water purification
E - Sand, water, cement

O Oxygen — O_2 gas; $O(2-,1-)$; ozone O_3 oxide $O(2-)$, peroxide $O_2(2-)$, hydroxide $OH(1-)$
C - **Permanent wave lotion**

17 — Halogens, X2; compounds: M(1-), Halide (1-), Acid=HX

At Astatine — metalloid
E, A, C - Seldom found in nature

I Iodine — solid; acid: HI, iodide$(1-)$, iodate $IO_3(1-)$, periodate $IO_4(1-)$
E - Disinfectant, halogen lamps
C - Ink pigments, salt additive

Br Bromine — liquid; acid: HBr, bromides $Br(1-)$, bromate $BrO_3(1-)$
C - Photographic film, tear gas
C - Fire retardants, disinfectant

Cl Chlorine — gas; Cl_2 gas; chlorate $ClO_3(1-)$ hypochlorite $ClO(1-)$, perchlorate $ClO_4(1-)$ acid: hydrochloric HCl, perchloric $HClO_4$, chloric $HClO_3$
E - Water purification
C - Bleach, hydrochloric acid, PVC plastics, stain removers

F Fluorine — F_2 gas; fluoride$(1-)$ acid: HF hydrofluoric, oxyfluorides-halofluorides
C - Toothpaste additive, teflon
C - Uranium enrichment
C - Refrigerator coolants

18 — Noble Gases; not reactive; unstable low temperature complexes

Rn Radon
E - Earthquake prediction, health threat in homes built on granite

Xe Xenon — Forms compounds with F
E - UV lamps, laser research
E - Photographic flashes, projection lamps, paint testers

Kr Krypton
E - **Fluorescent bulbs**, flashbulbs
E - UV lasers, wavelength standard

Ar Argon
E - Light bulbs, gas discharge tubes lasers
E - Geiger counters, welding blanket gas

Ne Neon
E - Neon lights, fog lamps, TV tubes
E - Lasers, voltage detectors

He Helium
E - Balloons, blimps, lasers
E - Diving bell atmosphere, leak detectors
E - Nuclear power plant coolant

MAJOR NATURAL ISOTOPES WITH % OF OCCURRENCE

Isotopes have the same number of protons, but different numbers of neutrons. Isotopes not listed are radioactive or synthetic.

- **H**: 1 - 99.98%; 2 - 0.02%
- **He**: 4 - ~100%
- **Li**: 6 - 7.5%; 7 - 92.5%
- **Be**: 9 - 100%
- **B**: 10 - 19.9%; 11 - 80.1%
- **C**: 12 - 98.9%; 13 - 1.1%
- **N**: 14 - 99.63%; 15 - 0.37%
- **O**: 16 - 99.8%; 18 - 0.2%
- **F**: 19 - 100%
- **Ne**: 20 - 90.5%; 21 - 0.3%; 22 - 9.3%
- **Na**: 23 - 100%
- **Mg**: 24 - 79.0%; 25 - 7.3%; 26 - 11.0%
- **Al**: 27 - 100%
- **Si**: 28 - 92.2%; 29 - 4.7%; 30 - 3.1%
- **P**: 31 - 100%
- **S**: 32 - 95.0%; 33 - 0.8%; 34 - 4.2%
- **Cl**: 35 - 75.8%; 37 - 24.2%
- **Ar**: 36 - 0.3%; 40 - 99.6%
- **K**: 39 - 93.3%; 41 - 6.7%
- **Ca**: 40 - 96.9%; 44 - 2.1%
- **Sc**: 45 - 100%
- **Ti**: 46 - 8.0%; 47 - 7.3%; 48 - 73.9%; 49 - 5.5%; 50 - 5.4%
- **V**: 50 - 0.3%; 51 - 99.7%
- **Cr**: 50 - 4.4%; 52 - 83.8%; 53 - 9.5%; 54 - 2.4%
- **Mn**: 55 - 100%
- **Fe**: 54 - 5.9%; 56 - 91.7%; 57 - 2.1%
- **Co**: 59 - 100%
- **Ni**: 58 - 68.1%; 60 - 26.2%; 61 - 1.1%; 62 - 3.6%; 64 - 0.9%
- **Cu**: 63 - 69.2%; 65 - 30.8%
- **Zn**: 64 - 48.6%; 66 - 27.9%; 67 - 4.1%; 68 - 18.8%
- **Ga**: 69 - 60.1%; 71 - 39.9%
- **Ge**: 70 - 21.2%; 72 - 27.7%; 73 - 7.7%; 74 - 35.9%; 76 - 7.4%
- **As**: 75 - 100%
- **Se**: 74 - 0.9%; 76 - 9.4%; 77 - 7.6%; 78 - 23.8%; 80 - 49.6%; 82 - 8.7%
- **Br**: 79 - 50.7%; 81 - 49.3%
- **Kr**: 80 - 2.3%; 82 - 11.6%; 83 - 11.5%; 84 - 57.0%; 86 - 17.3%
- **Rb**: 85 - 72.2%; 87 - 27.8%
- **Sr**: 86 - 9.9%; 87 - 7.0%; 88 - 82.6%
- **Y**: 89 - 100%
- **Zr**: 90 - 51.5%; 91 - 11.2%; 92 - 17.2%; 94 - 17.4%; 96 - 2.8%
- **Nb**: 93 - 100%
- **Mo**: 92 - 14.8%; 94 - 9.3%; 95 - 15.9%; 96 - 16.7%; 97 - 9.6%; 98 - 24.1%; 100 - 9.6%
- **Tc**: 93 - synthetic
- **Ru**: 96 - 5.5%; 98 - 1.9%; 99 - 12.7%; 100 - 12.6%; 101 - 17.1%; 102 - 31.6%; 104 - 18.6%
- **Rh**: 103 - 100%
- **Pd**: 102 - 1.0%; 104 - 11.1%; 105 - 22.3%; 106 - 27.3%; 108 - 26.5%; 110 - 11.7%
- **Ag**: 107 - 51.8%; 109 - 48.2%
- **Cd**: 106 - 0.9%; 108 - 0.9%; 110 - 12.5%; 111 - 12.8%; 112 - 24.1%; 113 - 12.2%; 114 - 28.7%; 116 - 7.5%
- **In**: 113 - 4.3%; 115 - 95.7%
- **Sn**: 112 - 1.0%; 116 - 14.5%; 118 - 24.2%; 119 - 8.6%; 120 - 32.6%; 122 - 4.6%; 124 - 5.8%
- **Sb**: 121 - 57.4%; 123 - 42.6%
- **Te**: 122 - 2.6%; 125 - 7.1%; 126 - 19.0%; 128 - 31.7%; 130 - 33.9%
- **I**: 127 - 100%
- **Xe**: 128 - 1.9%; 129 - 26.4%; 130 - 4.1%; 131 - 21.2%; 132 - 26.9%; 134 - 10.4%; 136 - 8.9%
- **Cs**: 133 - 100%
- **Ba**: 134 - 2.4%; 135 - 6.6%; 136 - 7.9%; 137 - 11.2%; 138 - 71.7%
- **La**: 138 - 0.1%; 139 - 99.9%
- **Ce**: 140 - 88.4%; 142 - 11.1%
- **Pr**: 141 - 100%
- **Nd**: 142 - 27.1%; 143 - 12.2%; 144 - 23.8%; 145 - 8.3%; 146 - 17.2%; 148 - 5.8%; 150 - 5.6%
- **Pm**: 145 - synthetic
- **Sm**: 144 - 3.1%; 147 - 15.0%; 148 - 11.3%; 149 - 13.8%; 150 - 7.4%; 152 - 26.7%; 154 - 22.7%
- **Eu**: 151 - 47.8%; 153 - 52.2%
- **Gd**: 154 - 2.2%; 155 - 14.8%; 156 - 20.5%; 157 - 15.7%; 158 - 24.8%; 160 - 21.9%
- **Tb**: 159 - 100%
- **Dy**: 160 - 2.3%; 161 - 18.9%; 162 - 25.5%; 163 - 24.9%; 164 - 28.2%
- **Ho**: 165 - 100%
- **Er**: 164 - 1.6%; 166 - 33.6%; 167 - 23.0%; 168 - 26.8%; 170 - 14.9%
- **Tm**: 169 - 100%
- **Yb**: 170 - 3.1%; 171 - 14.3%; 172 - 21.9%; 173 - 16.1%; 174 - 31.8%; 176 - 12.7%
- **Lu**: 175 - 97.4%; 176 - 2.6%
- **Hf**: 176 - 5.2%; 177 - 18.6%; 178 - 27.3%; 179 - 13.6%; 180 - 35.1%
- **Ta**: 180 - 0.01%; 181 - 99.99%
- **W**: 182 - 26.3%; 183 - 14.3%; 184 - 30.7%; 186 - 28.6%
- **Re**: 185 - 37.4%; 187 - 62.6%
- **Os**: 186 - 1.6%; 187 - 1.6%; 188 - 13.3%; 189 - 16.1%; 190 - 26.4%; 192 - 41.0%
- **Ir**: 191 - 37.3%; 193 - 62.7%
- **Pt**: 194 - 32.9%; 195 - 33.8%; 196 - 25.3%; 198 - 7.2%
- **Au**: 197 - 100%
- **Hg**: 198 - 10.0%; 199 - 16.9%; 200 - 23.1%; 201 - 13.2%; 202 - 29.8%; 204 - 6.9%
- **Tl**: 203 - 29.5%; 205 - 70.5%
- **Pb**: 204 - 1.4%; 206 - 24.1%; 207 - 22.1%; 208 - 52.4%
- **Bi**: 209 - 100%
- **La**: 138 - 0.1%; 139 - 99.9%
- **Th**: 232 - 100%
- **Pa**: 231 - 100%
- **U**: 235 - 0.7%; 238 - 99.3%

Elements >#92 are synthetic and radioactive